Luis Eduardo Ribeiro Valente
Andrécia Pereira da Costa
Juliete da Silva Souza

Characterizing the Profile of Photovoltaic Distributed Generation

Luis Eduardo Ribeiro Valente
Andrécia Pereira da Costa
Juliete da Silva Souza

Characterizing the Profile of Photovoltaic Distributed Generation

In the northern region of Brazil

ScienciaScripts

THANKS

To my father, Leonaldo Cardoso Valente, for his valuable advice and constant encouragement throughout this journey.

To my sister, Eduarda de Cássia Ribeiro Valente, for the emotional support and affection she gave me even at a distance.

To my partner, Jéssica dos Santos, for her constant presence and tireless support.

To my aunt and uncle, Lucivaldo and Odinéia, for welcoming me throughout my time in Tucuruí.

To my partner's family, especially my mother-in-law, Maria de Mercês, and my brother-in-law Jefferson dos Santos, for their welcome and wise advice.

To my friends Alessandro, Daniel, Gilzi, Jaciane and Camila, for sharing precious moments throughout the course.

To all the teachers, especially Jefferson Souza, André Cruz and Andrécia Pereira, for the knowledge they imparted during the course.

To my advisor Andrécia Pereira, for her guidance and all the time she made available.

To my co-supervisor Juliete da Silva, for her guidance and encouragement over the last few years.

And to everyone else who has contributed in some way to my academic career. Thank you for every gesture of help throughout the course.

"Let the future tell the truth and judge everyone according to their work and achievements.
The present belongs to them, but the future I've always worked for belongs to me."

(Nikola Tesla)

SUMMARY

This paper presents a study to characterize the profiles of photovoltaic distributed generation systems installed in the states that make up the Northern Region of Brazil, from 2013 to 2023. To analyze and extract data on the distributed generation profile of the Northern Region, the Sistema de Registro de Geração Distribuída (SISGD) platform on the website of the National Electric Energy Agency (ANEEL) was used. This online platform provides up-to-date information on distributed generation throughout the country and is accessible to the general public. The analysis of the results, using statistical data, identified a growth in the installation of distributed generation systems in Brazil, especially photovoltaic systems, which is mainly due to the government incentives provided for in Normative Resolutions No. 482/2012 and No. 687/2015. It was noted that in the context of the Northern Region, the state with the most photovoltaic generation systems was Pará. By understanding the global context and analyzing the specific results for the northern region of Brazil, this study contributes to understanding the ongoing energy transition, with a view not only to understanding the present, but also to projecting future prospects for solar energy generation in the region, considering its continued potential for expansion and contribution to a more sustainable and diversified energy matrix.

Keywords: Solar energy. Photovoltaic panels. Northern region. Energy transition. Sustainability.

LIST OF ABBREVIATIONS AND ACRONYMS

°C	Degrees Celcius
ANEEL	Agência Nacional de Energia Elétrica
DG	Distributed Generation
DG-PV	Photovoltaic Distributed Generation
GW	Gigawatt
GWth	Gigawatt Thermal
KW	Kilowatt
m-si	Monocrystalline silicon
p-si	Polycrystalline Silicon
SISGD	Distributed Generation Registration System
CUs	Consumer Units
W	Watt

CONTENTS

1. INTRODUCTION

Photovoltaic solar energy has emerged as a crucial catalyst in the transformation of the world's energy landscape, playing a significant role in the search for sustainable and clean sources. Growing awareness of the limitations of non-renewable resources and the impacts of greenhouse gas emissions make the switch to more sustainable sources vital. The incorporation of renewable energies not only reduces dependence on fossil fuels, but also contributes to mitigating climate change and building a more sustainable future (PINHO; GALDINO, 2014).

In this scenario, photovoltaic distributed generation systems (PV DG) have emerged as a promising and viable solution. The incentives for using these systems are not just limited to environmental aspects, but also offer significant advantages for consumers. Government support plays an essential role, and specific standards have been implemented to promote the integration of these technologies. Notable among these standards are National Electric Energy Agency (ANEEL) Normative Resolution No. 482/2012 and ANEEL Normative Resolution No. 687/2015 (ANEEL, 2024).

Normative Resolution 482/2012 establishes the general conditions for distributed micro and mini generation access to electricity distribution systems. Through it, consumers who opt for GD-FV systems can inject surplus energy into the electricity grid, obtaining credits to be used at times of lower generation, such as at night or during cloudy periods. Normative Resolution 687/2015 improves energy compensation mechanisms, simplifying the process of connecting to the grid and strengthening incentives for the implementation of these sustainable technologies. These ANEEL regulations not only reflect the Brazilian government's commitment to diversifying the energy matrix and promoting renewable sources, but also simplify procedures for consumers, encouraging the adoption of DG-PV and strengthening the transition to a more sustainable energy model (ANEEL, 2024).

The aim of this paper is to present a characterization study of the profiles of photovoltaic distributed generation systems installed in the states that make up the Northern Region of Brazil from 2013 to 2023, a total of 10 years of evolution in PV-DG in the region.

1.1 Justification

PV-DG has emerged as a viable alternative for promoting decentralized electricity generation, allowing consumers to generate some or all of the electricity they consume.

ANEEL's Normative Resolution No. 482/2012 and No. 687/2015 establish favorable conditions for integrating GD-FV systems into the electricity grid, simplifying things and offering incentives for consumers. The study aims to analyze the impact of these regulations on the increase in GD-FV installations in the Northern Region of Brazil.

1.2 Objectives

1.2.1 General objective:

To analyze the profile of photovoltaic distributed generation systems in the states of the Northern Region, taking into account different modalities, consumption class and growth in the period from 2013 to 2023.

1.2.2 Specific objectives:

The specific objectives of this work were:

- Study the main sources of renewable energy generation;
- Learn about the main aspects of distributed photovoltaic generation systems;
- Describe the characteristics of photovoltaic distributed generation systems in Brazil;
- To detail the characteristics of photovoltaic distributed generation systems in the northern states of Brazil;
- To evaluate the evolution of distributed generation systems in the northern region of Brazil.

1.3 Structure of the work

This work is divided into five chapters. Chapter 1 presents the introduction with the topic addressed, followed by the justification, finally establishing the general objective and specific objectives of the research. Chapter 2 presents the theoretical background, covering concepts about energy generated from solar sources, rules and regulations established by the government regarding the installation of photovoltaic panels and their connection types.

Chapter 3 details the methodology adopted in the study. Chapter 4 contains the results and discussions, highlighting data obtained from an ANEEL platform, which provides quantitative data on distributed generation. Finally, Chapter 5 presents the study's final considerations.

2. WORLD/NATIONAL ENERGY PANORAMA AND THE CONTRIBUTION OF PHOTOVOLTAIC GENERATION

Photovoltaic solar energy has emerged as one of the most promising solutions in the search for renewable energy sources. This paper sets out to explore the understanding of this specific form of energy generation, which uses photovoltaic cells to directly convert sunlight into electricity. Before delving into the specific details of photovoltaic solar energy, it is crucial to understand the broader panorama of the world's and Brazil's energy matrix, highlighting the growing importance of alternative sources in the global energy transition.

2.1 Electricity

In the study of electrical and magnetic phenomena, whose roots go back to Classical Antiquity, important milestones stand out that culminated in the scientific advances of the 18th and 19th centuries. Charles A. Coulomb, in 1784, was a pioneer in quantifying the force between charges using his torsion balance, revealing its inverse relationship with the square of the distance between them. Similarly, Carl F. Gauss, in 1813, made a significant contribution by demonstrating the relationship between the flow of the "electric force" and the electric charge inside a closed surface, while Hans C. Oersted, in 1820, surprised by observing the deflection of a magnetic needle by an electric current, highlighting the interconnection between electrical and magnetic phenomena (LAGE, 2021).

However, the most notable advances were driven by figures such as Michael Faraday, whose discoveries revolutionized the foundations of physics. In 1831, Faraday not only introduced the concept of the electric field as a condition of space generated by the presence of charges, but also formulated the law of induction, radically transforming the methods of generating and transmitting energy. James Clerk Maxwell, in 1864, consolidated this progress by establishing the fundamental equations of the electromagnetic field, demonstrating that disturbances in this field propagate at the speed of light in a vacuum, thus unifying optics with electromagnetism and paving the way for modern physics (LAGE, 2021).

2.2 Types of energy generation sources

Currently, the predominant energy matrix in much of the world is built around the use of fossil fuels, notably oil and natural gas, as its main sources of energy. This energy model, although widely established, presents a number of significant challenges that deserve critical attention. One of the predominant problems associated with this energy approach is the finiteness of the resources employed. In addition, the negative impacts on the environment resulting from the use of fossil fuels, such as atmospheric and water pollution, have become issues of concern and the subject of growing debate (SANTOS; RIBEIRO; MIRANDA, 2012).

Alternative energy represents a method of generating electricity that poses fewer challenges to contemporary society, has a lower environmental impact and results in a substantial reduction in the emission of pollutants (RAMOS, 2011).

Energy from renewable sources comes from natural resources, such as solar, wind, water, tidal and geothermal energy, which are naturally replenished (SANTOS; RIBEIRO; MIRANDA, 2012).

The distinction between renewable and non-renewable energy sources is of critical importance when analyzing the diversification and sustainability of a nation's energy matrix (IEA, 2021).

2.3 Electricity and energy matrices in Brazil and worldwide

The electricity matrix and the energy matrix are fundamental concepts in the study of energy consumption and distribution. The electricity matrix refers to the composition of electricity generation sources in a specific region, highlighting the proportion of energy produced from sources such as hydroelectric, wind and solar power plants, among others. The energy matrix, on the other hand, covers a broader sphere, considering not only electricity generation, but also the production of energy for transportation and heating, including the fuels used in vehicles and heating systems. This distinction is crucial for assessing the diversification of energy sources, sustainability and the environmental impacts associated with energy consumption (IEA, 2021).

The world's energy structure is predominantly made up of non-renewable energy sources, exemplified by the significant use of coal, oil and natural gas, as can be seen in Figure 1 (EPE, 2022).

Conventional energy sources such as oil derivatives, natural gas and coal make up around 80% of the world's energy matrix, while renewable energy sources such as solar, wind and geothermal collectively account for only 2.5% of the composition of the world's energy matrix, categorized as 'Other' in graphical representations. Added to the contribution of hydropower and biomass, renewable energy sources reach approximately 15% of the total composition Figure 1 (EPE, 2022).

Figure 1 - World energy matrix

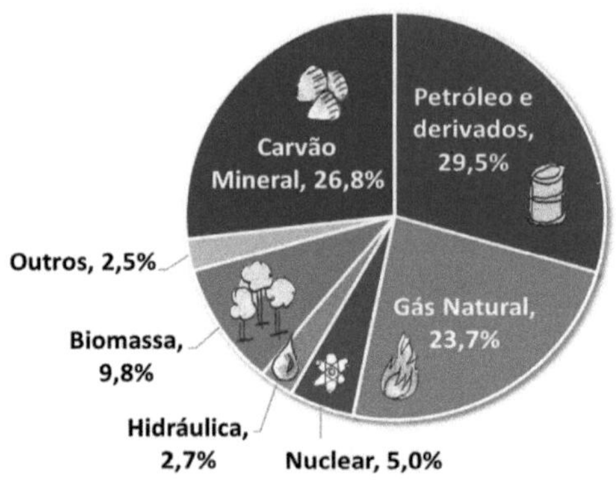

Source: EPE, 2022.

The composition of the Brazilian energy matrix differs substantially from the global configuration. In Brazil, there is a more significant predominance of renewable energy sources compared to the world average. When we consider the aggregation of resources such as firewood, charcoal, hydroelectric power, biofuels derived from sugar cane, wind power, solar power and other renewable sources, we see that energy from renewable sources accounts for 47.4%, representing almost half of the country's energy matrix. The second most used energy source in Brazil is oil, with over 35% of the national energy matrix, due to the oil wells found in the pre-salt, as can be seen in Figure 2 (EPE, 2022).

Figure 2 - Brazilian energy matrix

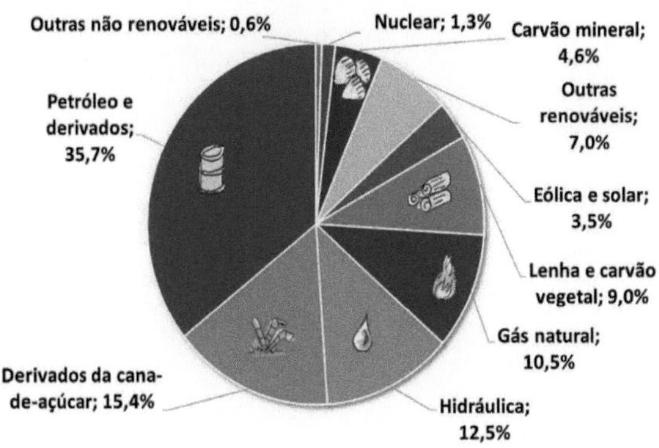

Source: EPE, 2022.

The electricity matrix represents the set of sources used exclusively to generate electricity at a national, regional or global level. Electricity is an urgent need, used in a variety of everyday contexts, such as television broadcasting, radio listening, domestic lighting, operating household appliances such as refrigerators, as well as charging mobile devices, among other essential applications (EPE, 2022).

Like the global energy matrix, the electricity matrix also shows a marked predominance in the use of conventional energy sources, representing more than 70% of the total composition. This scenario reflects a considerable dependence on non-renewable resources such as coal, oil and natural gas for electricity generation. Despite this, it is important to note that in recent years there has been a gradual increase in investment in and incorporation of renewable energy sources, such as solar, wind and hydroelectric, with a view to diversification and the search for more sustainable sources in electricity generation, as illustrated in Figure 3 (EPE, 2022).

Figure 3 - World electricity matrix

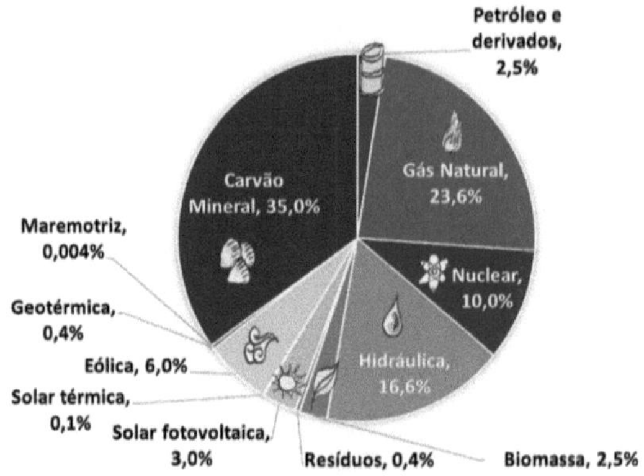

Source: EPE, 2022.

Brazil's electricity matrix stands out for its greater predominance of renewable sources compared to the energy matrix, driven significantly by the significant contribution of hydroelectric plants to electricity generation in the country, around 62%. In addition, the growing development of wind energy has played a fundamental role in the continued consolidation of the predominance of renewable sources in the Brazilian electricity matrix. It is also important to highlight the remarkable progress of solar energy, which has gradually increased its share, further strengthening the mostly renewable composition of our electricity matrix, as illustrated in Figure 4 (EPE, 2022).

Figure 4 - Brazil's electricity matrix

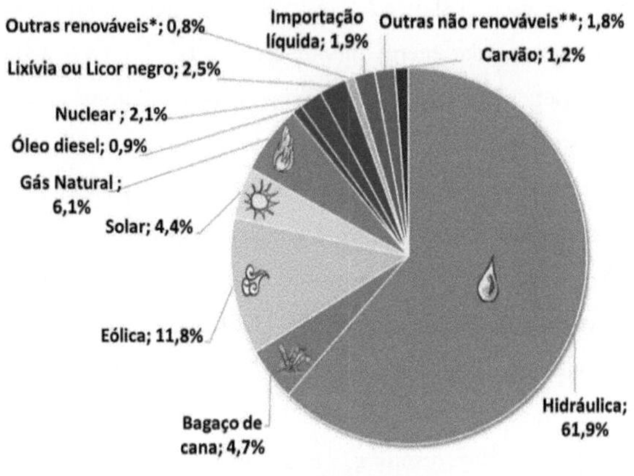

Source: EPE, 2022.

2.4 Solar power generation

Capturing solar radiation from the sun is one of the ways of obtaining electricity. The sun, a vast sphere of incandescent gas, is the primary source of this energy resource, generating energy through thermonuclear reactions (RAMPINELLI, 2010).

Solar radiation is a form of energy emitted by the sun, propagating in all directions of space in the form of electromagnetic waves. Its characteristics, such as its wide dispersion, result in low density and temporal variability. These attributes are fundamental and have substantial relevance for maximizing the use of available solar energy (RAMPINELLI, 2010).

2.4.1 Solar thermal energy

The main function of solar collectors is to heat fluids. They can be classified as concentrators or flat collectors, depending on the presence of devices to concentrate the solar radiation. The heated fluid can be stored in thermally insulated tanks until it is used. Concentrators are used in high-temperature applications that can reach up to 400°C to generate electricity using steam turbines. Flat collectors, on the other hand, are mainly used in low-temperature residential and commercial applications, such as heating water for bathing, drying grains, heating swimming pools, among other uses (PINHO; GALDINO, 2014).

The use of solar energy to heat water to temperatures below 100°C is currently one of the most common applications in Brazil. This practice eventually replaces electric or gas heating systems, which are used in showers. This preference stems from the simplicity of the technology for converting this solar energy into thermal energy, which is widely available on the market, offering a variety of suppliers and manufacturers. In addition, economic viability is easily achieved in well-structured projects (PEREIRA *et al.*, 2017).

In the residential sphere alone, expenditure on water heating accounts for around 24% of total electricity consumption. Electric showers, which can operate with powers of over 6000 W, contribute significantly to peak demand, which is generally concentrated at peak times, typically in the early evening, between 6pm and 9pm. This behavior increases the overall energy demand of the National Interconnected System (SIN), which leads to high costs in the expansion of generation, transmission and distribution systems, in order to meet high-power, low-use devices throughout the day, resulting in a low load factor (PEREIRA *et al.*, 2017).

The total capacity of solar heaters installed worldwide in 2016 exceeded 456 GWth (thermal gigawatt), corresponding to around 652 million square meters of installed collectors. China has emerged as the main market for new installations, with the use of evacuated tube collector technology predominating. Globally, in addition to evacuated tube collectors, conventional flat plate collectors with glass covers and polymeric or open collectors are widely used, as shown in Figure 5. Although Brazil has the third largest total installed capacity of solar heating systems, its per capita position is 30th in the *ranking*. This suggests ample room for growth in this market, especially when we consider that, in per capita terms, the country lags behind many others, even though it has a considerable availability of solar resources (PEREIRA *et al.*, 2017).

Figure 5 - The twenty largest total capacities of solar collectors in operation in 2016

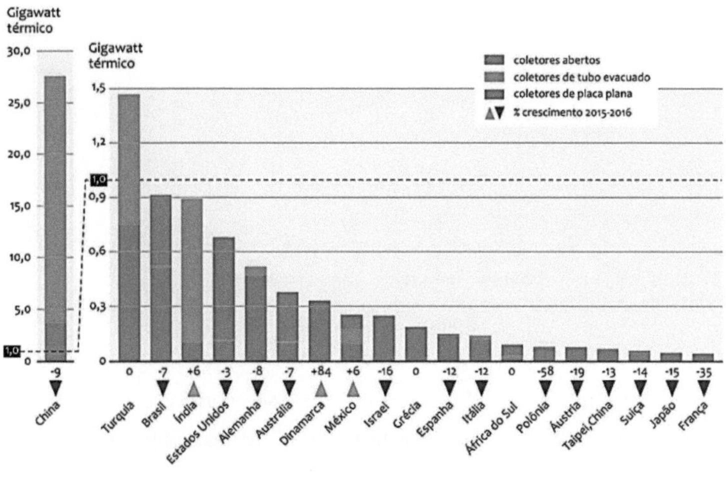

Source: Pereira *et al.*, 2017.

2.4.2 Photovoltaic solar energy

The phenomenon of the photovoltaic effect, originally identified by Edmond Becquerel in 1839, generates a potential difference at the terminals of an electrochemical cell when light is absorbed. In 1876, the first photovoltaic apparatus was developed from studies in solid state physics. It wasn't until 1956 that industrial production of these devices began, in line with the growth of the field of electronics (PINHO; GALDINO, 2014).

Photovoltaic solar energy is generated directly from sunlight, converting it into electricity through the photovoltaic effect. The primary unit in this process is the photovoltaic cell, made of semiconductor material. The technologies used to manufacture photovoltaic cells and modules are classified into three generations. The first generation includes monocrystalline silicon (m-Si) and polycrystalline silicon (p-Si), which dominate more than 85% of the market because they are considered consolidated, reliable technologies with the best commercial efficiency available (PINHO; GALDINO, 2014).

By the end of 2022, global photovoltaic solar energy capacity had reached 1,047 GW. There has been considerable annual growth in recent years, with an increase of 22.4% over the previous year and approximately six times more than in 2014. Between 2014 and 2022, this capacity grew at an annual average of 25.0%, as illustrated in Figure 6 (BEZERRA, 2023).

Figure 6 - Annual increase (%) and evolution of installed capacity of photovoltaic solar generation in the world (GW) - 2014 - 2022

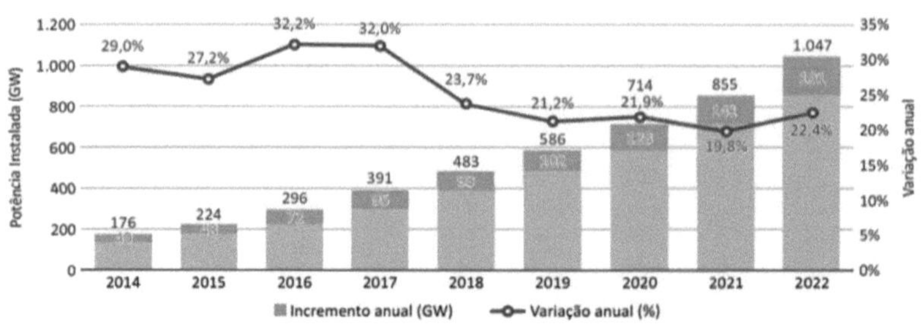

Source: IRENA (2023).

The outlook is that the remarkable growth in the production of photovoltaic solar energy will continue in the following years, due to the push for renewable energy sources and expectations of lower generation costs (BEZERRA, 2023).

2.5 Distributed generation in Brazil

Distributed generation, also known as dispersed, local or embedded generation, has as its main characteristic the production of electricity, which is carried out in the vicinity of or in conjunction with the consumer. This form of generation connects directly to the energy distribution networks through installation in consumer units (UCs) (INEE, 2020).

In Brazil, distributed generation is mentioned for the first time in Article 14 of Decree No. 5.163, of July 30, 2004. According to this definition, distributed generation refers to the production of electricity from projects by concessionaires, permit holders or authorized agents, including those referred to in Article 8 of Law No. 9,074 of 1995, and connected directly to the buyer's electrical distribution system. However, this definition excludes hydroelectric projects with an installed capacity of more than 30 MW and thermoelectric projects, including cogeneration, whose energy efficiency is less than seventy-five percent, according to ANEEL regulations established until December 2004 (BRASIL, 2004; INEE, 2020).

The term Distributed Generation (DG) was mentioned in 2004, but it wasn't until April 17, 2012 in Normative Resolution 482, issued by the National Electric Energy Agency, which is represented as a regulatory milestone, making it possible for consumers to exchange the energy they generate with the electricity grid. This resolution established the general conditions for distributed microgeneration and minigeneration access to electricity distribution systems, outlining the rules for offsetting credits, differentiating system power between consumer groups and setting the minimum rate applied to each consumer group (DUARTE, 2022).

2.6 Laws and decrees

Since ANEEL Normative Resolution 482/2012 came into force on April 17, 2012, Brazilian consumers have gained the right to generate their own electricity from renewable sources or qualified cogeneration. This right not only promotes self-sustainability, but also makes it possible to offset energy consumption. The Distributed Electricity Microgeneration and Minigeneration (MMGD) modalities and the Electricity Compensation System (SCEE) were developed with the aim of unifying economic benefits and socio-environmental awareness (ANEEL, 2024).

Over time, ANEEL has improved the MMGD rules, mainly through Normative Resolutions No. 687 (2015), No. 786 (2017) and, more recently, with Normative Resolution No. 1,059 of February 7, 2023. The aim of these changes is to bring the regulations into line with current legal requirements, consolidating the guidelines in Normative Resolution No. 1,000/2021 (ANEEL, 2024).

The rules allow the use of any renewable source, in addition to qualified cogeneration, with distributed microgeneration defined for up to 75 kW and distributed minigeneration for above 75 kW up to 3 MW (with the possibility of reaching 5 MW in specific situations). The SCEE makes it possible to inject surplus energy into the grid, functioning as a virtual battery to store and distribute the surplus as needed (ANEEL, 2024).

2.7 Shared generation

The National Electric Energy Agency (ANEEL) established the practice of shared generation through Normative Resolution 687/2015, which allows the sharing of energy from micro or mini-generation systems between two or more consumers (CPF or CNPJ) who are in the same concession area. This innovative form of energy generation implies compliance with specific guidelines, such as the formation of consortia or cooperatives and the need for a consumer unit with distributed micro or mini-generation (WEG, 2024).

In addition, it is crucial that the generation site is distinct from the one where the surplus energy will be compensated. Shared generation, operating through consortia or cooperatives, not only provides economic advantages, but also promotes sustainability by enabling the use of alternative sources such as solar energy. This cooperative approach stands out as an efficient alternative for both homes and businesses, allowing groups of consumers to optimize the use of renewable energy resources in a collaborative manner (WEG, 2024).

2.8. Remote self-consumption

ANEEL Resolution 687 of 2015 also introduced the definition of remote self-consumption as that which is characterized by consumer units owned by the same legal entity, including head offices and branches, or by an individual who owns a consumer unit with distributed microgeneration or minigeneration at a different location from their other consumer units. However, it is important to note that all these units must be located within the same concession or permission area. In this context, the surplus energy generated will be compensated in the system (ANEEL, 2024).

The applicant for the generating consumer unit must formalize the request for inclusion of the beneficiary consumer units by submitting the "form for registration of consumer units participating in the compensation system", duly completed and signed, together with the access request. Any changes can be requested at a later date, by submitting the same form at least 60 days prior to its implementation. The form must specify the percentage of the surplus to be transferred to each beneficiary (ANEEL, 2024).

In cases where there is a transfer of surplus, the unit where the generation plant will be installed will have its metering equipment replaced by one with bidirectional reading, while the other units will only record consumption and may keep their existing meters. It is important to note that bills remain individualized for each consumer unit (ANEEL, 2024).

2.9 Types of photovoltaic systems

The main feature of *on-grid* systems, also known as *grid-tie, is that they are* directly connected to the electricity grid, allowing voltage and current to be fed into it. These systems are recognized for complementing the Brazilian energy matrix, as they generally have a lower installed load than other energy sources, such as hydroelectric, wind or coal. They are usually installed in places that already have the infrastructure to connect to the power utility grid (ALVES, 2019).

Due to their ability to inject surplus energy into the grid, *on-grid* systems make it possible to apply the energy credit system. This is possible through the use of bidirectional energy meters, in which the utility company records both the energy consumed and the energy generated. This allows for compensation and the application of credits. Thus, the utility grid plays the role of a "virtual battery", absorbing the surplus energy generated by the *on-grid* system, illustrated in Figure 7 (ALVES, 2019).

Figure 7 - How an *on-grid* system works

Source: Luz Solar (2021).

The *on-grid* system illustrated in Figure 7 consists mainly of several photovoltaic panels connected to an inverter and then to the electricity grid. This system does not store energy, which means that the energy generated and not used by the consumer/generator is delivered directly to the electricity grid (ALVES, 2019).

Unlike *on-grid* systems, *off-grid* systems operate independently of the conventional grid. In this scenario, the system is responsible for supplying voltage and current to all the loads present in the installation, taking over the functions that would normally be performed by the electricity grid. This autonomy is made possible through the use of battery banks (ALVES, 2019).

Off-grid systems are prevalent in remote locations that lack electricity grid infrastructure. In these regions, the high costs associated with implementing infrastructure for energy distribution and transmission make it impractical to connect to the conventional grid. Therefore, the choice falls on the use of energy sources through grid-independent systems, as illustrated in Figure 8 (ALVES, 2019).

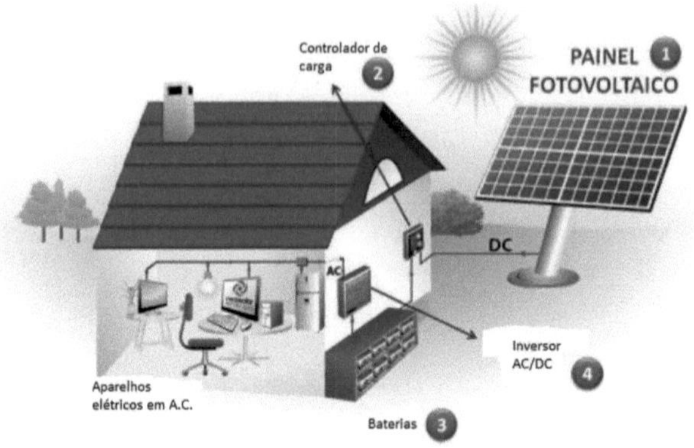

Source: Trevi Solar (2017).

In Figure 8, four components make up the *off-grid* photovoltaic system: the photovoltaic module, which is similar to *on-grid* systems; the charge controllers, responsible for protecting the battery against overcharging or discharging in smaller systems, disconnecting the photovoltaic generator when the battery reaches maximum charge; the batteries, common in isolated systems for storing energy, with lead-acid batteries predominating, despite other more efficient and durable options; and the inverter, similar to *on-grid* systems, which converts energy to power alternating current equipment, incorporating a maximum power point follower to optimize final production, and is used to provide comfort when using conventional household appliances (ALVES, 2019).

3. EXTRACTING DATA FROM THE SISGD PLATFORM

3.1 Characterization of the type of study

This study adopts an exploratory approach, focusing on the search for and compilation of bibliographic data in order to deepen our understanding of the subject in question. It is also characterized by a descriptive design, since the data presented is official and fundamental to the contextualization of this work.

The technical processes were conducted by means of bibliographical research and documentary analysis in official agency publications, as well as in specialized literature. This approach made it possible to obtain the qualitative and quantitative data needed to outline the profile of distributed photovoltaic generation in the northern region of Brazil.

3.2 Data collection and use system

This study was conducted by analyzing and extracting data available on the Distributed Generation Registration System (SISGD) platform. This online platform provides up-to-date information on distributed generation throughout the country and is accessible via the website of the National Electric Energy Agency (ANEEL). Using this platform makes it easy to search for information, allowing the use of search filters that provide accurate and transparent data (ANEEL, 2024).

In this work, search filters were used that included criteria such as generation source, region, states, generation type, consumption class, installed power and year of connection, established as January 1, 2013 to December 31, 2023.

In order to standardize the collection of information and thus avoid discrepancies in the final analysis resulting from possible updates in SISGD, the data was collected on January 24, 2024, between 7:00 pm and 10:45 pm.

4. RESULTS AND DISCUSSION

4.1 Overview of DG in Brazil

In order to analyze the Distributed Generation (DG) scenario in Brazil, as well as its geographical distribution and its various characteristics, we used data acquired from the ANEEL website. Figure 9 shows the regional distribution of DG in the country. It is worth noting that the data presented represents the total number of systems using wind, solar, hydroelectric and thermal resources.

Figure 9 - Data on the national distribution of DG systems (units)

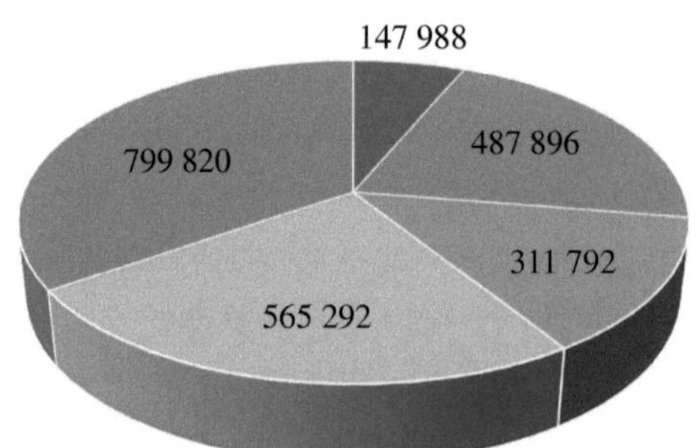

Source: Prepared according to (ANEEL, 2024).

Table 1 shows the number of systems for each type of renewable resource, broken down by region. The diversity in the distribution of systems for each resource in all regions stands out. Specifically in relation to photovoltaic (PV) generation, the Southeast stands out as the leader in terms of the number of systems, in contrast to the North, which has the lowest number of systems in this segment.

Table 1 - Data on DG in Brazil by Type and Region

Types of DG	Quantity of DG by Region				
	North	North East	South	South East	Midwest
Photovoltaic	147.987	487.830	565.039	799.499	311.699
CGH	0	1	18	49	8
Wind	4	42	37	13	1
Thermal	7	23	198	259	89

Source: Prepared according to (ANEEL, 2024).

When evaluating the systems that make use of small hydroelectric plants (CGHs), generally designed to take advantage of the hydroelectric potential of smaller rivers), it can be seen that the Southeast stands out with the largest number of installed systems, followed by the South, Midwest and Northeast, respectively. It is important to note that, in the context of these systems, the North Region did not have any systems registered during the period studied.

As far as DG systems from wind resources are concerned, the Northeast is the best-placed region, followed by the South, Southeast and North, respectively. In the Midwest, one installed wind generation unit was registered. With regard to systems based on thermal sources, it can be seen that the Southeast ranks first, preceded by the South, Midwest and North.

Considering the above, it is clear that DG systems using photovoltaic technologies stand out for their significant presence in all regions. In view of this finding, the subsequent analyses will be aimed at providing a more in-depth breakdown of the growth of these systems in the country and in the Northern Region, the focus of this research.

4.2 GD-PV scenario in Brazil

As seen in Table 1 (subtopic 4.1), Photovoltaic Distributed Generation (PV-DG) is the renewable source with the highest number of units installed in all regions of Brazil.

Figure 10 shows the installed power of PV DG systems in each of the country's states. It can be seen that the state of São Paulo is in first place, with an installed power of 3,510,588.39 kW, thus obtaining the highest index in the country, while the state of Roraima is in last place with an installed power of 37,479.24 kW. The sum of the installed power in Brazil was calculated by adding up all the figures for the Brazilian states in the five regions, giving a total of 25,903,813.90 kW of installed power in PV-GD systems in the country.

Figure 10 - Installed Power of DG - PV in the States of Brazil

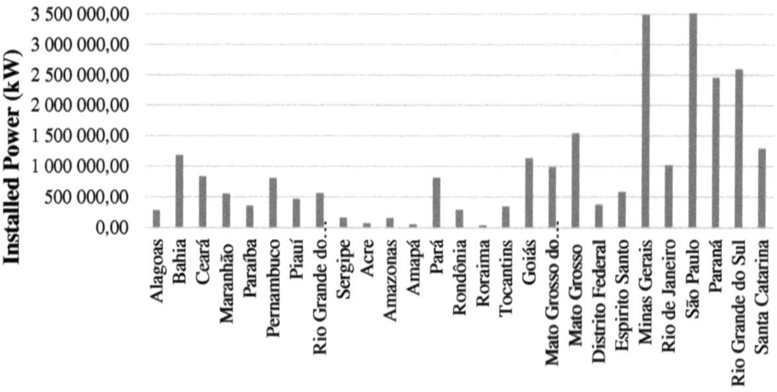

Source: Prepared according to (ANEEL, 2024).

Figure 11 shows the percentage of installed power in each of the country's regions in relation to the installed power in Brazil (25,903,813.90 kW). As can be seen, the Southeast has the highest percentage of installed power with 33.16% (8,590,568.56 kW), followed by the South with 24.47% (6,337,469.20 kW), the Northeast with 20.01% (5,184,042.46 kW), the Midwest with 15.57% (4,032,647.92 kW) and the North with 6.79% (1,759,085.76 kW).

Figure 11 - Installed Power of DG - PV in percentages by Regions of Brazil

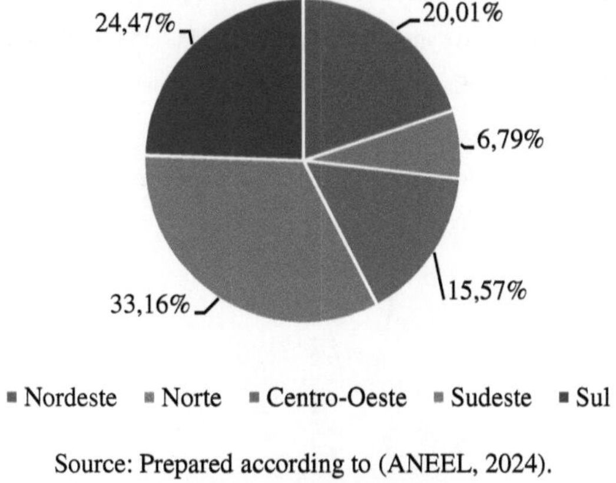

Source: Prepared according to (ANEEL, 2024).

We also analyzed the data on installed PV-GD power in each of the states in the country's five regions. Figure 12 shows the installed power in each state in the Southeast. It can be seen that São Paulo stands out with installed power of 3,510,588.39 kW equivalent to (40.87%), followed by Minas Gerais with 3,486,130.74 kW (40.58%), Rio de Janeiro with 1,016,175.39 kW (11.83%). Espírito Santo is the state with the least installed power, at 577,674.04 kW, equivalent to (6.72%) of the region's total power.

Figure 12 - Installed Power of DG - PV in percentages of the Southeast Region

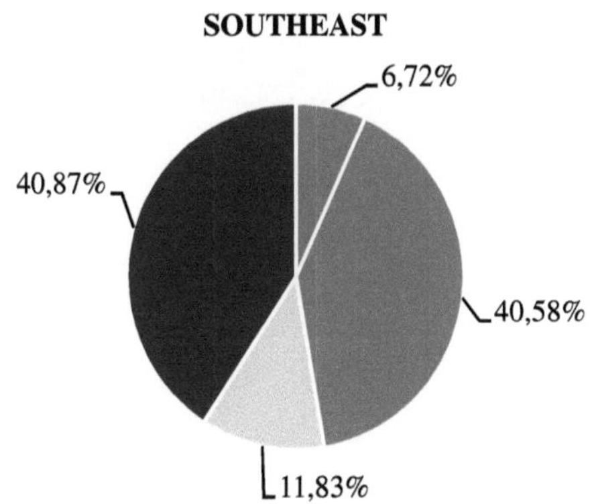

Figure 13 shows the data obtained for PV-GD in the South. The state of Rio Grande do Sul stands out with installed power of 2,594,165.14 kW, equivalent to 40.93%, while the states of Paraná and Santa Catarina have installed power of 2,454,355.65 kW and 1,288,948.41 kW, equivalent to 38.73% and 20.34% respectively of the total power in the Southern Region.

Figure 13 - Installed Power of DG - PV in percentages of the South Region

SOUTH

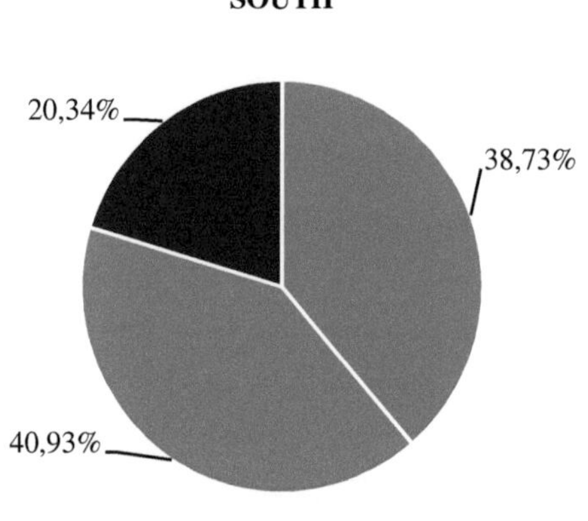

Source: Prepared according to (ANEEL, 2024).

Figure 14 shows the installed power in the Midwest region. The state of Mato Grosso stands out with 1,538,167.42 kW (38.14%), followed by Goiás with 1,134,528.42 kW (28.13%) and Mato Grosso do Sul with 985,572.94 kW (24.44%). The Federal District has the lowest installed capacity, 374,379.14 kW (9.28%).

Figure 14 - Installed Power of DG - PV in percentages of the Center-West Region

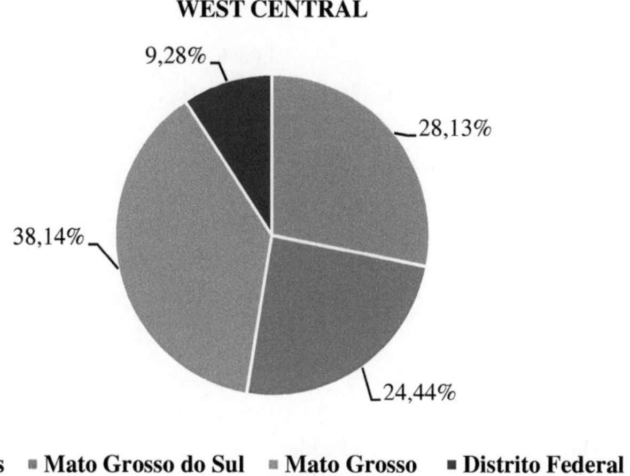

Source: Prepared according to (ANEEL, 2024).

Figure 15 shows the percentage of installed power in the Northeast Region, which is led by Bahia with 1,178,703.22 kW (22.74%), followed by Ceará 836,014.88 kW (16.13%), Pernambuco 804.467.04 kW (15.52%), Rio Grande do Norte 560,657.15 kW (10.82%), Maranhão 549,347.01 kW (10.60%), Piauí 461,731.92 kW (8.91%), Paraíba 355,003.07 kW (6.85%), Alagoas 278,571.32 kW (5.37%) and Sergipe with 159,546.85 kW (3.08%), respectively.

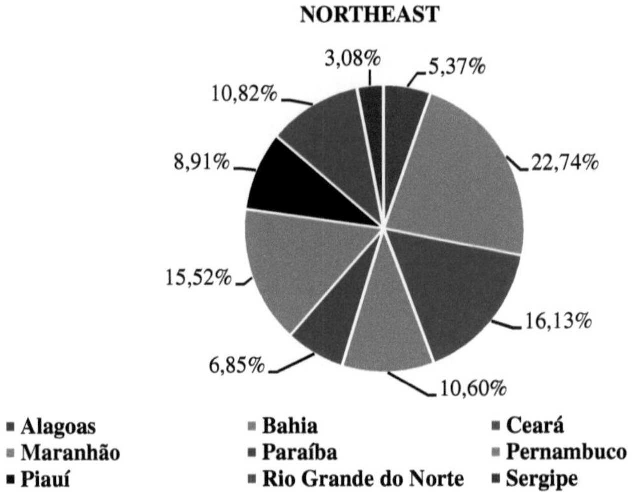

Source: Prepared according to (ANEEL, 2024).

The data for the Northern Region, the focus of this study, can be seen in Figure 16. It can be seen that the state of Pará stands out with 810,347.49 kW (46.23%), followed by Tocantins with 343,421.00 kW (19.59%), Rondônia with 282,822.38 kW (16.14%), Amazonas with 158.559.48 kW (9.05%), Acre with 68,766.50 kW (3.92%), Amapá with 51,430.45 kW (2.93%) and Roraima with 37,479.24 kW (2.14%) of the installed power in the north of the country.

Figure 16 - Installed Power of DG - PV in percentages of the North Region

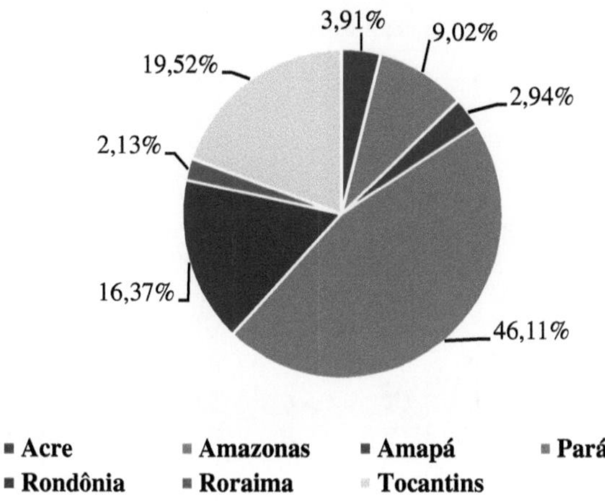

Source: Prepared according to (ANEEL, 2024).

4.3 Analysis of PV-GD in the northern region of Brazil

In order to provide a more in-depth understanding, this section will explore the characteristics of Photovoltaic Distributed Generation (PV-DG) systems, taking into account the type of generation, consumption class and the evolution of installed power in each state of the Northern Region.

4.3.1 Profile of DG-PV by generation type

The panorama of photovoltaic generation, categorized by modality, reveals that Remote Self-Consumption in the North Region has a total of 22,475 systems. In this context, the state of Pará stands out, having installed 13,127 systems, while Roraima has 211 systems.

In the Shared Generation category, the region is home to 181 systems, 71 of which are located in the state of Pará, while Amapá has 1 installed system and the state of Acre has no records. As for Own Generation, the Northern Region has 125,325 units of installed systems, with special emphasis on the state of Pará, which contributes 58,130 systems, while Roraima has 1,807 installed systems.

As for the Multiple HU modality, the North Region has a total of 6 systems, with 5 of them located in Pará and 1 in Rondônia; there are no records in the other states. These figures can be seen in Figure 17.

Figure 17 - Number of systems by type of generation in the North Region

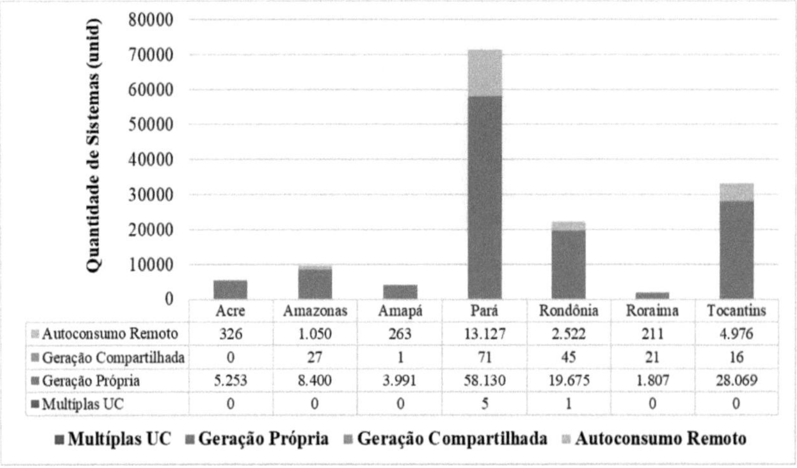

	Acre	Amazonas	Amapá	Pará	Rondônia	Roraima	Tocantins
Autoconsumo Remoto	326	1.050	263	13.127	2.522	211	4.976
Geração Compartilhada	0	27	1	71	45	21	16
Geração Própria	5.253	8.400	3.991	58.130	19.675	1.807	28.069
Multiplas UC	0	0	0	5	1	0	0

Source: Prepared according to (ANEEL, 2024).

Figure 18 illustrates the number of Consumer Units (CUs) benefiting from the generation of energy from PV-GD systems in the Northern Region, categorized according to the type of generation. In the case of remote self-consumption, the state of Pará stands out with 33,591 CUs benefiting from this energy source, while the state of Amapá has 719 CUs benefiting.

Figure 18 - Number of CUs by generation type in the North Region

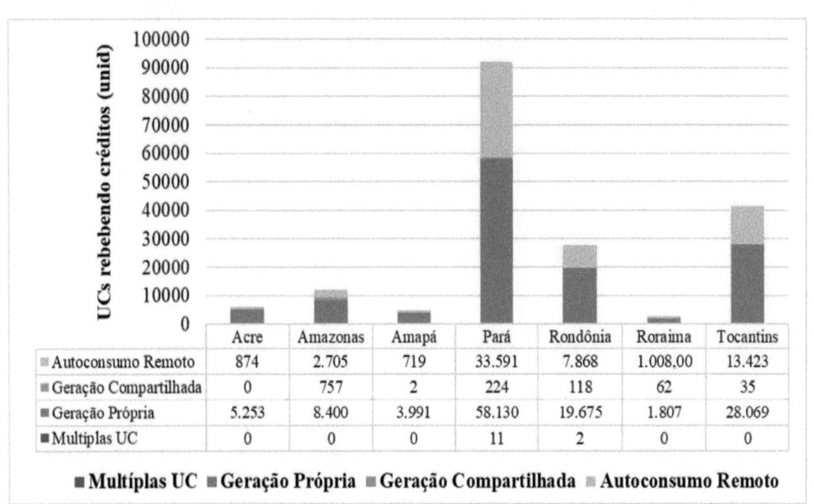

	Acre	Amazonas	Amapá	Pará	Rondônia	Roraima	Tocantins
Autoconsumo Remoto	874	2.705	719	33.591	7.868	1.008,00	13.423
Geração Compartilhada	0	757	2	224	118	62	35
Geração Própria	5.253	8.400	3.991	58.130	19.675	1.807	28.069
Multíplas UC	0	0	0	11	2	0	0

Source: Prepared according to (ANEEL, 2024).

In the context of self-generation, the number of installed systems coincides with the number of Consumer Units (CUs) in all the states of the Northern Region.

In the Shared Generation category, the state of Amazonas stands out with 757 Consumer Units (CUs), while the state of Amapá has only 1 system installed, benefiting 2 CUs. In the multiple CU mode, the state of Pará stands out with 5 systems installed, benefiting 11 CUs. On the other hand, the states of Acre, Amazonas, Amapá, Roraima and Tocantins do not have any systems installed, resulting in a lack of Consumer Units benefiting from this type of generation.

Figure 19 shows the installed power in each state of the Northern Region, according to the type of generation. The following observations stand out: In the own generation modality, the state of Pará leads with an installed power of 602,689.66 kW, while Roraima records a power of only 23,676.63 kW. In remote self-consumption, the state of Pará stands out with 203,975.00 kW of installed power, in contrast to the state of Acre, which has only 6,256.62 kW. In shared generation, the state of Pará has 4,357.33 kW of installed power, while Amapá has the lowest installed power in the region in this modality, with just 5 kW.

As for the Multiple UC modality, the state of Pará has 5 systems installed, totaling 76.5 kW, and Rondônia has one system with 15 kW of installed power. In the states of Acre, Amazonas, Amapá, Roraima and Tocantins, there are no systems installed, resulting in no data on installed power in this generation modality.

Figure 19 - Installed capacity by type of generation in the North Region

	Acre	Amazonas	Amapá	Pará	Rondônia	Roraima	Tocantins
Autoconsumo Remoto	6.256,62	25.050,04	8.025,30	203.975,00	46.724,64	12.455,55	69.556,55
Geração Compartilhada	0	3.902,12	5	4.357,33	932,16	1.347,06	365,7
Geração Própria	62.558	129.674,32	43.721,39	602.689,66	240.222,06	23.676,63	273.498,75
Multíplas UC	0	0	0	76,5	15	0	0

Source: Prepared according to (ANEEL, 2024).

4.3.2 Profile of PV-GD by consumption class

Figure 20 shows the distribution of installed power in each state of the Northern Region, considering consumption classes. The consumption classes are classified as follows: commercial, public service, rural, residential, public power, industrial and public lighting.

Figure 20 - Installed Power by Consumption Class in the Northern Region

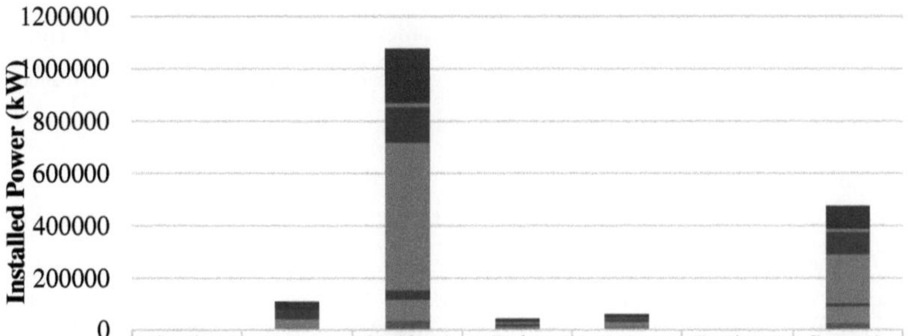

	Serviço Público	Rural	Residencial	Poder público	Industrial	Iluminação Pública	Comercial
■ Tocantins	3 010,00	26 509,23	206 309,42	10 363,54	11 252,07	39,00	85 937,74
■ Roraima	0	767,19	15 501,70	8 177,37	789,00	10,00	12 233,98
■ Rondônia	0	38 468,90	137 784,65	9 992,90	15 646,43	10,00	85 990,98
■ Pará	10	36 311,68	563 873,63	5 761,07	19 081,22	60,00	186 000,89
■ Amapá	60	47,99	35 425,65	2 124,89	402,84	0	13 690,32
■ Amazonas	70	1 853,74	81 550,87	2 294,13	9 238,07	0,00	63 619,67
■ Acre	12	3 372,23	35 739,81	2 277,85	1 783,28	0,00	25 629,83

■ Acre ■ Amazonas ■ Amapá ■ Pará ■ Rondônia ■ Roraima ■ Tocantins

Source: Prepared according to (ANEEL, 2024).

In the commercial and residential consumption classes, the state with the highest installed power is Pará, while in these two consumption classes the state of Roraima has the lowest installed power in the region. In rural consumption, the state with the highest installed power is Rondônia (38,468.90 kW) and the lowest is Amapá (47.99 kW).

In the public service class, the state of Tocantins has the highest installed capacity in the region at 3,010.00 kW, while Pará has an installed capacity of 10 kW and the states of Roraima and Rondônia have no installed capacity. In the public power class, the state of Tocantins has the highest installed power of 10,363.54 kW, while the state of Amapá has the lowest installed power in this class at 2,124.89 kW.

In the Industrial class, Pará has 19,081.22 kW, the largest installed capacity, while Amapá has an installed capacity of 402.84 kW. In the public lighting class, Pará has the highest installed power of 60 kW, while the states of Roraima and Rondônia have 10 kW of installed power and the states of Amapá, Amazonas and Tocantins have no systems installed in this consumption class.

During the analysis, there was a significant variation in the installed power, the number of systems and the number of Consumer Units (CUs) receiving credits from PV-GD energy generation in the consumption classes studied. In order to facilitate data analysis, an average of the installed power per consumer unit (CU) was made, allowing the identification of the consumption classes that stand out in each location, as illustrated in Figure 21.

Figure 21 - Average installed power per HU by consumption class in the North Region

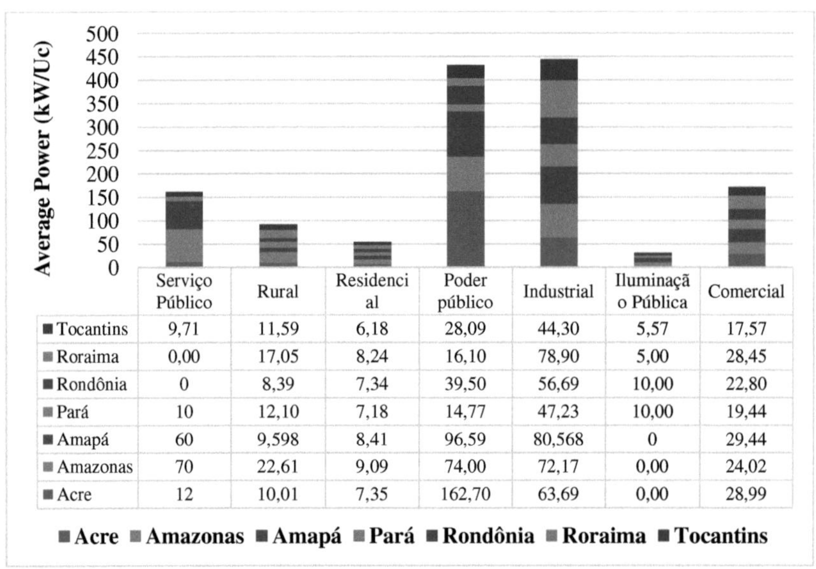

	Serviço Público	Rural	Residencial	Poder público	Industrial	Iluminação Pública	Comercial
■ Tocantins	9,71	11,59	6,18	28,09	44,30	5,57	17,57
■ Roraima	0,00	17,05	8,24	16,10	78,90	5,00	28,45
■ Rondônia	0	8,39	7,34	39,50	56,69	10,00	22,80
■ Pará	10	12,10	7,18	14,77	47,23	10,00	19,44
■ Amapá	60	9,598	8,41	96,59	80,568	0	29,44
■ Amazonas	70	22,61	9,09	74,00	72,17	0,00	24,02
■ Acre	12	10,01	7,35	162,70	63,69	0,00	28,99

Source: Prepared according to (ANEEL, 2024).

It can be seen that in the Industrial class, most states in the North have the highest average power per HU, while the public lighting class has the lowest ratio of power per HU, with the states of Amapá, Amazonas and Acre having no installed power at all.

It can be seen that in the public power class, it is the second class to stand out in most of the states of the Northern Region. The highest average power index recorded was in the state of Acre with 162.70 kW/UC and the lowest was in the state of Pará with 14.77 kW/UC.

The Commercial class was third, with the highest average power value recorded in the state of Acre with 28.99 kW/UC and the lowest in the state of Tocantins with 17.57 kW/UC. Next came the Rural class with the highest average installed power in Amazonas with 22.61 kW/UC and the lowest in the state of Rondônia with 8.39 kW/UC.

As for the Public Service class, the states of Amazonas and Amapá stand out with the highest average installed power with 70 kW/UC and 60 kW/UC, respectively. The states of Roraima and Rondônia have no systems in this consumption class.

In general, the residential class had a low rate of installed power per HU, where the highest value of power per HU recorded was 9.09 kW/UC in the state of Amazonas and the lowest was in the state of Tocantins with 6.18 kW/UC.

By way of comparison, Figure 22 shows a comparison of the average power per HU in the North Region and in Brazil. Similar trends can be seen when considering the classes with the highest and lowest average powers, with a divergence in the public lighting and public service classes. The industrial class has the highest average installed power per UC (North: 52.66 kW/UC and Brazil: 28.11 kW/UC) and residential the lowest (North: 7.14 kW/UC and Brazil: 5.43 kW/UC).

Figure 22 - Comparison of average installed power in the North and Brazil

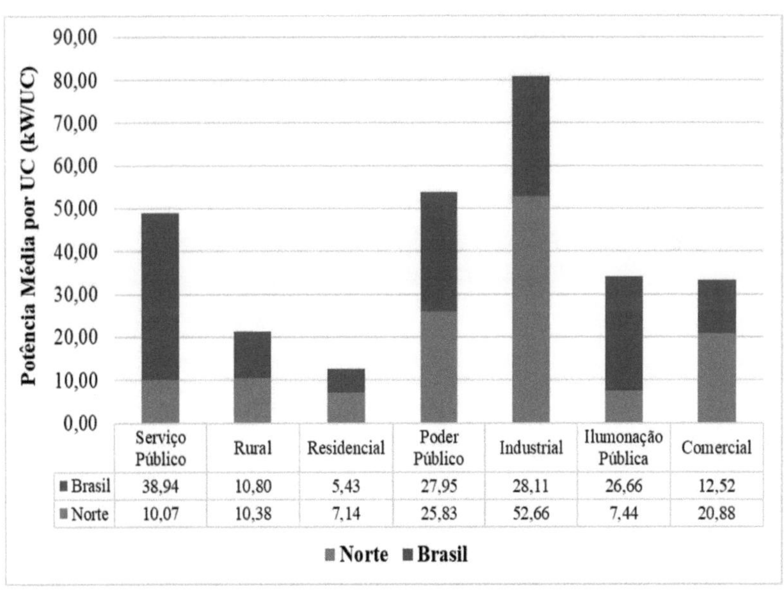

	Serviço Público	Rural	Residencial	Poder Público	Industrial	Ilumonação Pública	Comercial
Brasil	38,94	10,80	5,43	27,95	28,11	26,66	12,52
Norte	10,07	10,38	7,14	25,83	52,66	7,44	20,88

Source: Prepared according to (ANEEL, 2024).

4.3.3 GD-PV profile by year of connection

This analysis shows the evolution over time of the number of new connections per year of PV-GD systems in each northern state. It is worth noting that the start of this process was observed from 2013 until 2023. Figure 23 shows a steady increase in the number of connections over the years. The state of Roraima had no new connections to PV-GD systems in 2013 and 2014. The state of Amapá had no new connections between 2013 and 2016.

The state of Pará stands out in terms of the number of new connections between 2013 and 2023, surpassed in 2015 and 2017 by the state of Tocantins.

Figure 23 - Evolution of installed power in the Northern Region

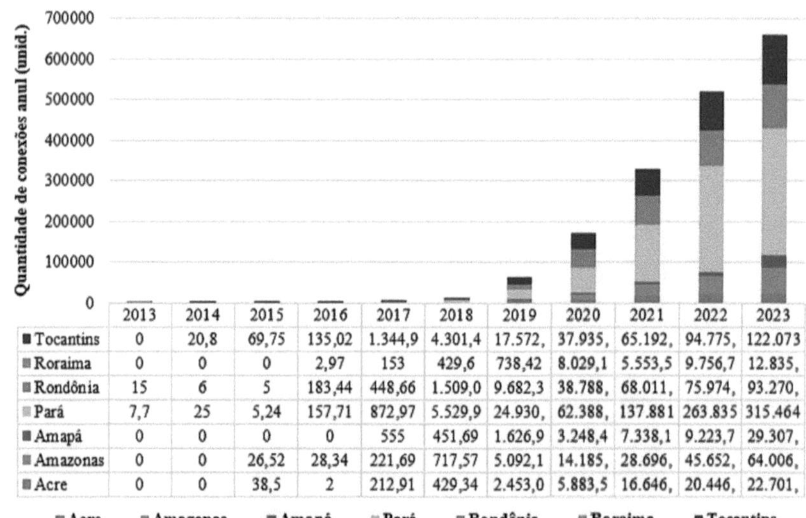

	2013	2014	2015	2016	2017	2018	2019	2020	2021	2022	2023
▪ Tocantins	0	20,8	69,75	135,02	1.344,9	4.301,4	17.572,	37.935,	65.192,	94.775,	122.073
▪ Roraima	0	0	0	2,97	153	429,6	738,42	8.029,1	5.553,5	9.756,7	12.835,
▪ Rondônia	15	6	5	183,44	448,66	1.509,0	9.682,3	38.788,	68.011,	75.974,	93.270,
▪ Pará	7,7	25	5,24	157,71	872,97	5.529,9	24.930,	62.388,	137.881	263.835	315.464
▪ Amapá	0	0	0	0	555	451,69	1.626,9	3.248,4	7.338,1	9.223,7	29.307,
▪ Amazonas	0	0	26,52	28,34	221,69	717,57	5.092,1	14.185,	28.696,	45.652,	64.006,
▪ Acre	0	0	38,5	2	212,91	429,34	2.453,0	5.883,5	16.646,	20.446,	22.701,

Source: Prepared according to (ANEEL, 2024).

Looking at the period from 2020 to 2022, it can be seen that most of the states in the North region experienced an increase in connections, coinciding with the COVID-19 pandemic scenario. This data shows a notable expansion of renewable energies, especially photovoltaic solar energy, in the region.

In this context, various sectors have adopted the implementation of photovoltaic projects as an effective strategy to mitigate the costs associated with electricity. This trend towards the adoption of photovoltaic generation systems has intensified over the last five years, gaining relevance especially during the COVID-19 pandemic. During this crisis, the increase in electricity costs has motivated consumers, especially in the industrial, commercial, rural and residential classes, among others, to look for alternatives to reduce these expenses, driving the adoption of sustainable solutions, mainly due to the attractive credit compensation system established in Resolutions 482 and 687.

This system gives owners of photovoltaic energy systems connected to the electricity grid the ability to generate energy credits whenever they produce a surplus in relation to their consumption. These credits, in turn, can be used to offset electricity consumption at times when solar generation is not sufficient, such as at night or on cloudy days. The result is a substantial reduction in energy bills and, in some cases, the possibility of receiving financial compensation, contributing not only to economic efficiency, but also to sustainability and energy autonomy for consumers.

5. FINAL CONSIDERATIONS

Solar radiation stands out as a renewable source that offers renewable energy without harming the environment. The sun plays a fundamental role in the evaporation of water, contributing to the formation of river reservoirs and enabling the generation of energy in hydroelectric plants. In addition, atmospheric circulation driven by solar radiation makes it possible to capture wind energy in wind farms. This sustainable approach underscores the importance of solar energy as a fundamental resource for electricity production, making efficient use of natural elements.

In Brazil's energy matrix, renewable sources are the most widely used, accounting for just over 70% in 2022. Of particular note in this context is the use of hydroelectric power generation.

However, the country has other resources available, such as high levels of solar radiation and a high wind regime. These resources are used in the country's electricity matrix. With regard to DG systems, this paper presents the distribution of systems throughout the country. Most of these systems are of the PV type, which justifies the focus adopted in this study. The country's Southeast region has the highest installed capacity (39.49%), while the North has the lowest installed capacity (7.03%).

This case study made it possible to identify the characteristics of PV-GD systems in the Northern Region by classifying them according to the type of generation, consumption class and the time evolution of the number of new installations. In the course of the research, it was noted that Pará was the state in the North Region that stood out with the most installed power, and the consumption class with the largest share in the state is Residential with 69.51% and around 563,873.63 kW of installed power. In terms of generation, the Own Generation modality has 602,689.66 kW of installed power distributed in 58,130 systems and benefiting 58,130 UCs.

The evolution over time of these systems shows an increase in the number of new system connections in the North, with the first registration of these PV-GD systems in the states of Pará and Rondônia in 2013. It wasn't until 2017 that PV-GD systems were registered in all the states, with Pará having the highest number of new connections between 2013 and 2018, but being surpassed in 2015 and 2017 by the state of Tocantins, although in the years after 2018, there was exponential growth in the state of Pará.

In view of this, the following are suggestions for future work:

- Study of the temporal evolution in the characteristics of GD-FV systems in the North;
- Study of the market potential of PV-GV systems belonging to the Industrial class in the North.

REFERENCES

NATIONAL ELECTRICITY AGENCY. **List of Distributed Generation projects**. 2024. Available at: https://dadosabertos.aneel.gov.br/dataset/relacao-de-empreendimentos-de-geracao-distribuida Distributed. Accessed on: January 24, 2024.

ALVES, M. O. L. **Solar energy: study of electricity generation through on-grid and off-grid photovoltaic systems**. 2019. Monograph (Undergraduate degree in Electrical Engineering) - Institute of Exact and Applied Sciences, Federal University of Ouro Preto, João Monlevade, 2019. Accessed on: January 23, 2024.

ANEEL. **Distributed Micro and Mini-generation**. Available at: https://www.gov.br/aneel/pt-br/assuntos/geracao-distribuida. Accessed on: January 22, 2024.

BEZERRA, Francisco Diniz. **Infrastructure:** Solar Energy. Fortaleza: BNB, year 8, n. 295, jul. 2023.(Caderno Setorial Etene).

BRAZIL. DECREE NO. 5.163 OF JULY 30, 2004. **Presidency of the Republic - Civil House**, 2004. Available at: http://www.planalto.gov.br/ccivil_03/_ato2004-2006/2004/decreto/D5163.htm. Accessed on: November 30, 2023.

BONJORNO, José Roberto *et al*. **Fundamental Physics**: single volume. São Paulo: Ftd, 1999.

DUARTE, Flávia Victória Souto. DISTRIBUTED GENERATION FOR MULTIPLE CONSUMER UNITS AND DESIGN OF A PHOTOVOLTAIC SYSTEM FOR A CONDOMINIUM. 2022. 85 f. TCC (Graduation) - Electrical Engineering Course, Federal University of Santa Maria, Santa Maria, 2022. Available at: http://repositorio.ufsm.br/handle/1/23780. Accessed on: January 25, 2024.

EPE. **Energy and Electricity Matrix**, 2022. Available at: https://www.epe.gov.br/pt/abcdenergia/matriz-energetica-e-eletrica. Accessed on: November 30, 2023.

IEA. **Key World Energy Statistics**, 2021. Available at: https://www.iea.org/reports/key-world-energy-statistics-2021. Accessed on: November 30, 2023.

IRENA. **Renewable capacity statistics**, 2023. Available at: <https://www.irena.org/Publications/2023/Mar/Renewable-capacity-statistics-2023>. Accessed on: November 29, 2023.

INEE. What is Distributed Generation. **INEE - National Institute for Energy Efficiency**, 2020. Available at: http://www.inee.org.br/forum_ger_distrib.asp#:~:text=Gera%C3%A7%C3%A3o%20Distribu %C3%ADda%20(GD)%20%C3%A9%20uma,Co%2Dgeradores/. Accessed on: November 30, 2023.

LAGE, E., **The fundamentals of electromagnetism,** Rev. Ciência Elem., V9(1):016, 2021. Available at: http://doi.org/10.24927/rce2021.016

Sunlight. **How photovoltaic systems work,** 2021. Available at: https://luzsolar.com.br/como-funciona-o-sistema-fotovoltaico/. Accessed on: December 25, 2023.

PEREIRA, E. B.; MARTINS, F. R.; GONÇALVES, A. R.; COSTA, R. S.; LIMA, F. L.; RÜTHER, R.; ABREU, S. L.; TIEPOLO, G. M.; PEREIRA, S. V.; SOUZA, J. G. **Atlas brasileiro de energia solar.** 2.ed. São José dos Campos: INPE, 2017. 80p. Available at: http://doi.org/10.34024/978851700089

PINHO, João Tavares; GALDINO, Marco Antonio. **Engineering Manual for Photovoltaic Systems**. 2. ed. Rio de Janeiro: Cepel, 2014.

RAMALHO, Francisco Junior *et al*. **The Fundamentals of Physics 1**: mechanics. 9. ed. São Paulo: Moderna, 2007.

RAMPINELLI, Giuliano Arns. **STUDY OF THE ELECTRICAL AND THERMAL CHARACTERISTICS OF INVERTERS FOR GRID-CONNECTED PHOTOVOLTAIC SYSTEMS**. 2010. 254 f. Thesis (Doctorate) - Mechanical Engineering Course, Federal University of Rio Grande do Sul, Porto Alegre, 2010. Available at: http://www.lume.ufrgs.br/bitstream/handle/10183/27935/000764646.pdf?sequence=1. Accessed on: November 17, 2023.

SANTOS, L. G. P., RIBEIRO, I. M., & MIRANDA, E. C. F. (2012). **Alternative energy**: the bridge to the future. Bolsista de valor: Revista de divulgação do projeto Universidade Petrobras e IF Fluminense, 261-269.

TREVI SOLAR. **OFF-GRID PHOTOVOLTAIC SYSTEM**, 2017. Available at: https://consultrevisolar.com.br/sistema-fotovoltaico-off-grid/. Accessed on: February 1, 2024.

WEG. **How does shared energy generation work and what are its benefits?** Available at: https://www.weg.net/solar/blog/como-funciona-a-geracao-compartilhada-de-energia-e-quais-sao-seus-beneficios/. Accessed on: January 21, 2024.

Printed by Books on Demand GmbH, Norderstedt / Germany